Seaweeds As Vegetables

Roby Jose Ciju

Copyright © 2014 AGRIHORTICO

All rights reserved.

Contents

1 SUPERFOOD SEAWEEDS 1

2 FOOD SIGNIFICANCE OF SEAWEEDS 3

3 SEAWEEDS AN INTRODUCTION 12

4 NUTRIENTS IN SEAWEEDS 14

5 RED SEAWEED GRACILARIA 17

6 IRISH MOSS 20

7 WINGED KELP 22

8 GREEN LAVER 24

9 PURPLE LAVER 26

10 SPIRULINA 27

11 WAKAME 29

12 DULSE 32

13 CAULERPA 33

14 KOMBU 34

15 GROWING SEAWEEDS 35

16 HEALTH BENEFITS OF SEAWEEDS 37

1 SUPERFOOD SEAWEEDS

Seaweeds are a group of edible algae. They are cholesterol-free, caffeine-free, low-fat, low-sugar foods that are rich in vitamins, minerals, protein and dietary fibres and therefore, considered as superfoods. Superfoods are multifunctional foods that contribute towards consumer's health and wellness. They are rich in antioxidants and are consumed to prevent modern-day life style diseases such as heart attack, diabetes, cancer and osteoporosis. Seaweeds are rich in antioxidants that help us to stay fit and young. Antioxidants are present in seaweeds in the form of vitamins, minerals, enzymes and polyphenolic compounds.

Major antioxidant vitamins are Vitamin C and Vitamin E. Major antioxidant minerals are Zinc, Manganese and Selenium. Selenium is essential to form an active site of most antioxidant enzymes. Manganese is required by the body as a co-factor for the antioxidant enzyme, *superoxide dismutase*. Three groups of antioxidant enzymes present in human body are *Superoxide Dismutase* (SOD), *Catalase (CAT) and Glutathione* and *Glutathione Peroxidise*. All the three are working together to protect body cells from free radical damage.

Polyphenolic compounds or polyphenols are a large group of antioxidants comprising of flavonoids, carotenoids, and anthocyanins. Flavonoids are chemical compounds that plants produce to protect themselves from cell damage. Flavonoids are a subgroup of polyphenolic antioxidants. It reduces cell inflammation, improves memory and concentration, and increases body's immunity. Anthocyanins are a subgroup of flavonoids. In case of carotenoids, there are about 600 types of carotenoids known till date. Some of these are alpha carotene, beta carotene, lycopene, cryptoxanthin, zeaxanthin, and lutein. Beta carotene is the most studied carotenoids and is a precursor of Vitamin A.

Health through Antioxidants

An antioxidant is a substance that inhibits oxidation, especially that of free radicals. Free radicals are chemically unstable molecular fragments or atoms that have a charge due to excess or deficient number of electrons and are directly responsible for cell degeneration and resultant ageing process in human beings. The immediate tendency of free radicals, as soon as they are formed, is to become stable by reacting with cellular components (for example: DNA) or cell membrane. The result is DNA damage, malignant tumour formation (cancer), diabetes, cataract, heart diseases and other cell degenerative diseases.

Examples of free radicals are superoxide anion, hydroxyl radical, transition metals such as iron and copper, nitric acid and ozone. Major sources of free radicals are normal oxidation process happening within the human body (i.e. released as a byproduct of cell metabolism), exposure to pollution (free radicals may be present in the air we breathe), exposure to sunlight and lifestyle factors such as alcohol consumption, wrong diet habits (free radicals may be present in the food we eat), stress, and smoking. Examples of cell damage caused by free radicals are cataract (lens of the eyes become opaque), damage to cell's protective lipid layer (cell membrane), and heart diseases where free radicals trap LDL (low density lipoprotein) in blood artery walls and form coatings. *When antioxidants are present in human body, they work towards protecting the body from cell damage by preventing the formation of free radicals.*

2 FOOD SIGNIFICANCE OF SEAWEEDS

Chlorophylls, the basic photosynthetic units, are present in most of the algae and they are able to synthesize their own food. They are excellent sources of protein, dietary fiber, vitamins and minerals. A detailed account of nutrition present in seaweeds is given below:

Water Content in Seaweeds

Skin needs to be well hydrated all the times to keep its healthy appearance. Our bodies are composed of up to 70% of water. Blood, brain, muscles, lungs etc all contain a lot of water. Body needs water to regulate the body temperature and for transporting nutrients, removing bodily toxins and waste, and for protecting body organs. A dehydrated body accelerates the ageing process. When body is dehydrated it results in energy loss, headaches and fatigue. We need to consume lots of water every day. Water also acts as appetite suppressant and hence helps in losing body weight. Freshly harvested seaweeds are rich in moisture. On an average, they contain up to 80 to 90 % water, soon after harvest. The table below shows moisture presence in 100 grams of edible portion of fresh seaweeds.

Water in Seaweeds (per 100 g edible portion)

Nutrient	Water
Unit	g
Agar seaweed	91.32
Irishmoss seaweed	81.34
Kelp seaweed	81.58
Laver seaweed	85.03
Spirulina seaweed	90.67
Wakame seaweed	79.99

Energy Value of Seaweeds

Fresh seaweeds are low-calorie foods making it ideal for weight management and anti-obesity treatments. A right diet should be high in nutritional value and low in calories. The food must provide sufficient calories of energy for bodily functions but not more. All foods, whether carbohydrates, fats or proteins, contain calories, i.e. provide energy. Calorie is a unit of measure of energy. A successful diet program contains right measures of energy. It is all about balance: balance between the calories consumed and the calories required by a body to perform its functions. Surplus calorie consumption results in overweight and insufficient consumption of calorie results in underweight. Both the extremes need to be avoided while planning a balanced diet. The table below shows energy value provided by 100 grams of edible portion of fresh seaweeds.

Energy in Seaweeds (per 100 g edible portion)

Nutrient	Energy
Unit	kcal
Agar seaweed	26
Irishmoss seaweed	49
Kelp seaweed	43
Laver seaweed	35
Spirulina seaweed	26
Wakame seaweed	45

Protein, Fat, Carbohydrate, Fiber and Sugar in Seaweeds

Protein is essential for body's tissue building and muscle generation. Protein metabolism releases amino acids, basic building blocks of a cell. *Seaweeds are excellent sources of good quality protein which is easily digestible and available to the human body.*

Carbohydrates are a vital component of a balanced diet and are a major source of energy. While sufficient carbohydrates must be present in a right diet, its surplus must be avoided. Carbohydrate overload increases bodily stress and as a result occurrence of life style diseases such as type 2 diabetes and heart ailments may be enhanced. *Seaweeds are low in carbohydrates thus making it ideal for a low-carb diet.*

Dietary fiber is good for human body. S*eaweeds are moderate source of dietary fiber.* High fiber foods is good for weight loss as they take long time to get digested and also make you feel full for a long period of time. High fiber food aids in digestion, cures constipation, lowers blood cholesterol, cleanses the gastrointestinal tract and may reduce the risk of developing diabetes and colorectal cancer. Two types of dietary fibers are found in vegetables: soluble and insoluble. Soluble fibers are dissolved in water while insoluble fibers are not dissolved. A soluble fiber helps in lowering blood cholesterol level as well as glucose level. Insoluble fibers help in movement of body toxins and waste through the digestive system. Insoluble fibers also relieve the body from constipation. S*eaweeds are mostly soluble fiber.*

Fats are a group of chemical compounds present in the human body. There are saturated fats and unsaturated fats. Saturated fat is 'bad' fat while unsaturated fat is 'good' fat. Saturated fats are mostly animal-origin while unsaturated fats are mostly plant-origin. Consuming unsaturated fats lower blood cholesterol while saturated fats increase blood cholesterol and heart ailments. S*eaweeds are normally very low in total lipid (fat) content.*

A right diet avoids foods high in unhealthy sugars. A human body is unable to break down large amounts of sugars and as a result additional sugars get deposited in the body tissues and body ages rapidly. It is always advisable to eat less sugar. S*eaweeds are low in sugar content.*

Protein, Fat, Carbohydrate, Fiber and Sugar in Seaweeds (per 100 g edible portion)

Nutrient	Protein	Fat	Carb	Fiber	Sugars
Unit in Grams (g)					
Agar Seaweed	0.54	0.03	6.75	0.5	0.28
Irishmoss	1.51	0.16	12.29	1.3	0.61
Kelp	1.68	0.56	9.57	1.3	0.6
Laver	5.81	0.28	5.11	0.3	0.49
Spirulina	5.92	0.39	2.42	0.4	0.3
Wakame	3.03	0.64	9.14	0.5	0.65

Minerals in Seaweeds

Minerals are essential elements that are required by a human body in minute quantities to perform its metabolic functions as well as for healthy growth and development. Minerals such as potassium, calcium, sodium, iron, magnesium, phosphorus, chlorine etc help in body fluid movement and tissue building. When a human body does not have sufficient amount of minerals, its deficiency is manifested in human body as diseases.

Calcium (Ca): Major functions of Ca are blood clotting, nerve impulse and muscle contraction, and protection of bones from thinning. Calcium promotes bone health and teeth health. Calcium deficiency results in weakening of bones, rickets in children, tooth decay and pains in legs and back of the body. *Seaweeds are excellent source of calcium.*

Magnesium (Mg): Magnesium promotes appetite and regulates blood pressure. It also promotes kidney health. Magnesium deficiency may result in poor appetite, weakness, osteoporosis, kidney stones, depression and anxiety, and high blood pressure. *Seaweeds are excellent source of magnesium.*

Phosphorous (P): Phosphorus is an essential component for the formation of high energy compounds and various nucleic acids. Phosphorous is a major component of bone and teeth and it also increases body's immunity. Phosphorous deficiency may result in anaemic appearances, weaknesses in muscles, poor immune symptoms etc. *Seaweeds are excellent source of phosphorous.* However, some seaweeds such as agar seaweeds and spirulina are not rich in phosphorus.

Iron (Fe): Iron is essential for RBC (red blood cell) formation and transfer oxygen between the tissues and within the blood. Red colour of blood is due to the presence of iron in it. Iron is essential for increasing body immunity and also for the synthesis of neurotransmitters. Iron deficiency results in anaemic appearances, increase in heart rate and breathing, poor immune system, yellowing of the body and severe headaches. *Seaweeds are good sources of iron.*

Zinc (Zn): Zinc strengthens body immune system, helps in clotting of blood and for the synthesis and digestion of proteins. It also helps in regulation of blood sugar and cholesterol. Zinc deficiency results in diarrhoea, loss of appetite, rashes on skin, weakness in sensing, night blindness, weak immune system etc. *Zinc is present in seaweeds in adequate amounts.*

Sodium (Na): Our major intake of sodium is through the consumption of salt. A healthy diet avoids too much consumption of salt. Salt is sodium chloride. Sodium presence in salt is calculated by using the formula salt=sodium x 2.5. Too much consumption of sodium increases blood pressure. It is high potassium to sodium ration which is critical in maintaining healthy blood pressure levels. *Since seaweeds are grown in sea, they are generally high in sodium content.*

Potassium (K): Potassium is an essential mineral that plays an important role in lowering blood pressure. *Seaweeds contain adequate amounts of potassium.*

Iodine (I): Iodine is essential for the synthesis of thyroid hormone and protects against goiter. Iodine also acts as an anti-oxidant. Iodine deficiency results in reduced production of thyroid hormone resulting in goiter, fatigue, reduced growth and increase in weight. *Sea weeds are excellent sources of iodine.*

Copper (Cu): Copper helps in absorbing iron from the gastrointestinal tract. Copper is essential for RBC (red blood cell) and WBC (white blood cell) formation. It is also a powerful antioxidant mineral. It is a major component of enzymes. Copper deficiency results in decrease in RBCs and WBCs, improper skin pigmentation and growth impairment in children. *Some seaweeds are believed to contain adequate amounts of copper.*

Manganese (Mn): Manganese is essential for the formation of amino acids, enzyme activation, formation of bones and functioning of muscles and nervous system. *Some seaweeds are believed to contain adequate amounts of manganese.*

A detailed account of minerals present in seaweeds is given below:

Minerals in Seaweeds (per 100 g edible portion)

Nutrient	Ca	Fe	Mg	P	K	Na	Zn
Unit in Milligrams (mg)							
Agar seaweed	54	1.86	67	5	226	9	0.58
Irishmoss	72	8.9	144	157	63	67	1.95
Kelp	168	2.85	121	42	89	233	1.23
Laver	70	1.8	2	58	356	48	1.05
Spirulina	12	2.79	19	11	127	98	0.2
Wakame	150	2.18	107	80	50	872	0.38

Vitamins in Seaweeds

We need vitamins in small quantities for healthy living and staying fit. Vitamins are required to regulate body metabolism and tissue building. Requirement of vitamins increases with the increase in age. Insufficient amount of vitamins in the body results in vitamin deficiency diseases. There are fat-soluble and water-soluble vitamins.

Fat-Soluble Vitamins

Vitamin A, D, E, and K (ADEK) are fat-soluble vitamins. That means, surplus consumption of these vitamins gets deposited in body fat and therefore daily consumption of these vitamins are not required.

Vitamin A: Vitamin A is also known as Retinol. It is essential for eye health. It also strengthens body's natural immune system. Vitamin A is also essential for tissue building, and skin health. Vitamin A deficiency results in night blindness, and drying of skin and eyes. S*eaweeds are excellent source of Vitamin A.* An exception is agar seaweeds in which Vitamin A is not present.

Vitamin D: Vitamin D is essential for bone health. Its deficiency results in rickets which is weakening of bones in children and softening of bones in adults. The deficiency also results in osteoporosis and muscle weakening in adults. Generally, Vitamin D is stored in human body while the body is exposed to sun light. Vitamin D is not present in seaweeds.

Vitamin E (alpha-tocopherol): Vitamin E is essential for strengthening body's natural immune system and cardiovascular system. It is a powerful antioxidant vitamin and hence protects the body from heart diseases and cancer. Vitamin E deficiency results in weakening of muscular system and nervous system. Other deficiency symptoms include lack of coordination and balance. S*eaweeds contain adequate amounts of Vitamin E.*

Vitamin K (phylloquinone): Vitamin K is essential for blood clotting, and for preventing heart diseases, cancer, and osteoporosis. Vitamin K deficiency results in bleeding gums and bleeding nose. *Seaweeds are excellent source of Vitamin K.*

A detailed account of fat-soluble vitamins present in seaweeds is given below:

Vitamins ADEK in Seaweeds (per 100 grams of edible portion)

Nutrient	Vit A	Vit D	Vit E	Vit K
Unit	IU	IU	mg	Âμg
Agar seaweed	0	0	0.87	2.3
Irishmoss	118	0	0.87	5
Kelp	116	0	0.87	66
Laver	5202	0	1	4
Spirulina	56	0	0.49	2.5
Wakame	360	0	1	5.3

Water-Soluble Vitamins

Vitamin B and C are known as water-soluble vitamins. Body cannot store water soluble vitamins such as Vitamin B and Vitamin C and therefore any surplus amount is eliminated from body through urine. Daily consumption of these vitamins is essential to stay healthy and young. Water-soluble vitamins may get destroyed while cooking. Hence vegetables containing Vitamin B and Vitamin C must be cooked by steaming or grilling rather than by boiling or deep frying.

Vitamin B Complex

Vitamin B complex contains Vitamin B1 (thiamine), B2 (riboflavin), B3 (Niacin or Nicotinic acid), B5 (Pantothenic acid), B6 (Pyridoxine), B7 (biotin), B9 (Folate/Folic acid) and B12 (Cobalamin).

Vitamin B1 or Thiamin: Vitamin B1 is also known as Thiamine (thiamin). It is essential for proper functioning of muscular and nervous systems. It also facilitates fatty acid production in the body and is essential for energy production within the body. Its deficiency disorder is called *Beriberi*, major symptoms of which is improper functioning of muscular and nervous systems. Generally, *seaweeds contain adequate amounts of thiamin required by the human body*. Spirulina seaweed is an excellent source of thiamin.

Vitamin B2 or Riboflavin: Vitamin B2 is also called Riboflavin. It is essential for eye health, skin health, hair health and energy metabolism. It is a powerful antioxidant vitamin. It also helps in the activation of Vitamin B6 and Vitamin B4. Major deficiency symptoms include swelling and redness of mouth, lips, tongue and skin. Another deficiency symptom is anaemia due the decreased RBC (red blood cell) count. *Seaweeds are excellent source of riboflavin.*

Vitamin B3 or Niacin: Vitamin B3 is also called Niacin or Nicotinic acid. It is essential for skin health, proper functioning of nerves, and digestion. It also reduces blood cholesterol level and therefore risk of heart attack. Deficiency disorder is called *Pellagra*. Deficiency symptoms include rashes on the skin, dementia and diarrhoea. The more severe case of the deficiency leads to death. *In seaweeds, niacin is present in small quantities.*

Vitamin B5 or Pantothenic acid: Vitamin B5 is also called Pantothenic acid. It is essential for proper functioning of adrenal gland, and carbohydrate metabolism. Vitamin B5 also helps in the production of bile acids, RBCs, cholesterol, fat and hormones. Deficiency symptoms include fatigue, headache and burning sensation in hands and feet. *Among the seaweeds, kelps are good source of Vitamin B5.*

Vitamin B6 or Pyridoxine: Vitamin B6 is also known as Pyridoxine. It is essential for fat metabolism and protein metabolism. It also helps in the production of RBCs and neurotransmitters. Vitamin B6 facilitates proper functioning of estrogen and testosterone hormones in the body. Deficiency symptoms include depression, improper functioning of immune system and sores in mouth. *Vitamin B6 is present in seaweeds in small quantities.*

Vitamin B7 or Biotin: Vitamin B7 is also called Biotin or Vitamin H. It helps in the synthesis of amino acids, glycogen as well as fat fibers. It is important for replication of DNA and maintaining better health of nails and hair. Deficiency symptoms include red rashes on eyes and in mouth and improper functioning of immune system. *Seaweeds contain adequate amounts of biotin required by human body.*

Vitamin B9 or Folate: It is also called Folic acid or Folate. It is essential for energy production from food. It helps in synthesis of nucleic acids and proper functioning of immune system and blood production by facilitating functioning of iron and increasing production of RBCs. It also helps in controlling amino acid metabolism. Major deficiency symptoms include birth defects in new born babies, diarrhoea, hearing loss due to ageing, improper functioning of immune system, weakness, fatigue and headaches. Regular consumption of folic acid helps in slowing down progression of hearing loss with ageing; to prevent birth related defects in new born babies; for protection from cancer, heart diseases, depression and degeneration of body due to ageing; and to prevent memory loss and osteoporosis. *Seaweeds are excellent source of folate.*

Vitamin B12 or Cobalamin: Vitamin B12 is also called Cobalamin. It helps in the synthesis of nucleic acids (DNA and RNA), RBCs, proper brain functioning, and energy metabolism. Deficiency symptoms include loss of appetite, anaemia, constipation, and depression. It is sometimes used as a remedy for asthma, male infertility, heart disorders and cancer. Cobalamin is found naturally in animal foods. Seaweeds are poor sources of Cobalamin.

Vitamin C: Vitamin C is also known as ascorbic acid. It is a powerful antioxidant vitamin. Vitamin C helps in absorption of iron and calcium. It increases body's natural immunity. Vitamin C deficiency results in a disease called *scurvy*. Major symptoms of scurvy are bleeding gum, joint pain, and hair loss. *Seaweeds are poor source of Vitamin C.*

Vitamins B and C in Seaweeds (per 100 grams of edible portion)

Nutrient	Thiamin	Riboflavin	Niacin	Vit B-6	Folate	Vit C
Unit	mg	mg	mg	mg	µg	mg
Agar	0.005	0.022	0.055	0.032	85	0
Irishmoss	0.015	0.466	0.593	0.069	182	3
Kelp	0.05	0.15	0.47	0.002	180	3
Laver	0.098	0.446	1.47	0.159	146	39
Spirulina	0.222	0.342	1.196	0.034	9	0.9
Wakame	0.06	0.23	1.6	0.002	196	3

3 SEAWEEDS AN INTRODUCTION

Seaweeds or sea vegetables are a group of marine macroalgae that are found growing on the rock surfaces or other hard surfaces along the coastal areas and sea bottoms. They belong to three different groups, on the basis of thallus (algal body) colour: brown algae (e.g. kelps), red algae (e.g. Gelidium) and green algae (e.g. green laver). Brown and red algae are marine in nature and therefore they are very large seaweeds while green algae are common in fresh waters and are comparatively small in size.

Brown colour of the brown algae is due to the presence of *fucoxanthin*, a xanthophyll pigment. Red colour of the red algae is due to the presence of pigments called phycoerythrin and phycocyanin. Green colour of the green algae is due to the presence of chlorophylls just like the higher plants. Food reserves in brown algae are polysaccharides, sugars, and higher alcohols. In red algae, food reserves are in the form of floridean starch and floridoside. Green algae reserves food as starch just like higher plants. In brown algae, cell walls are made up of cellulose and alginic acid. Cell walls in red algae are made up of cellulose and complex polysaccharides such as agars and carrageenans. In green algae, cell walls are made up of cellulose.

Seaweeds are now used as multifunctional foods in many parts of the world. Seaweeds are considered as functional foods because of the presence of high amounts of minerals and vitamins as well as some beneficial bioactive compounds in them. A functional food produces a beneficial effect in human body by increasing the welfare and decreasing the risk of diseases. When it comes to the matter of health, *prevention is better than cure* and functional foods are preventative than curative.

Seaweeds are suitable for mass production owing to the fact that they are autotrophic, marine macroalgae and require only the water-dissolved carbon dioxide and nutrients for its growth. They require no or little external resources for its mass production. Despite these positive factors, the commercial cultivation of seaweeds as food is not yet picked up in many countries. Currently, Asian countries, particularly China, Japan and Korea dominate the global seaweed production. France, UK, Chile, Philippines, Indonesia, Norway, USA, Canada and Ireland are also producing seaweeds in considerable quantities.

Seaweeds are produced for two major food purposes: for direct human consumption and for processing for further food applications such as stabilizing, gelling and thickening agents to be used in food and confectionery industry.

Seaweeds are directly consumed as raw (e.g. as salads and sandwich ingredients) or cooked form such as pickles, and vegetables. Dried and powdered seaweeds are used as seasoning and food flavoring agents. Jelly derived from certain seaweeds is used as thickening agent in ice creams, soups, marmalades etc. Phycocolloids such as alginates, agar and carrageenan present in seaweeds are extracted and used for food processing purposes.

Seaweeds are used for preparing various nutraceuticals or food supplements. Food supplements have important health benefits. Certain bioactive compounds derived from seaweeds are used as therapeutic agents in pharmaceutical industry. Seaweeds are also used as organic manure, compost, and as a meal for poultry and cattle. Seaweeds are used for the extraction of industrial gums and chemicals and also for manufacturing cosmetics.

With the advent of organic food and organic farming revolution, seaweeds are now increasingly becoming popular as a naturally-grown vegetable. It is also gaining popularity the world over as an organic super food.

4 NUTRIENTS IN SEAWEEDS

All edible seaweeds are cholesterol-free and caffeine-free and are considered as healthy foods. They are high in minerals iodine and calcium and contain soluble dietary fiber. Edible seaweeds are rich source of high quality protein. Some of the seaweeds contain considerable amounts of antioxidants such as fucoxanthin, fucosterol, and phlorotannin. Since seaweeds contain considerable amounts of fibre, protein, minerals, vitamins and low fat carbohydrate content, they are the best choice for a balanced diet. Nutritional information of popular six seaweeds such as *gracilaria spp.*, irish moss, winged kelp, green laver, spirulina and wakame is briefed below:

Minerals in Edible Seaweeds

Major minerals present in edible seaweeds are calcium, iron, magnesium, phosphorus, potassium, sodium, zinc, and iodine.

Mineral Composition of Seaweeds per 100g Edible Portion (unit in milligrams)

Seaweed	Ca	Fe	Mg	P	K	Na	Zn
Gracilaria spp.	54	1.86	67	5	226	9	0.58
Irish Moss	72	8.9	144	157	63	67	1.95
Kelp	168	2.85	121	42	89	233	1.23
Laver	70	1.8	2	58	356	48	1.05
Spirulina	12	2.79	19	11	127	98	0.2
Wakame	150	2.18	107	80	50	872	0.38

Source: USDA database

All these seaweeds are rich source of iron, potassium and zinc. 150 milligrams or more calcium is present in kelp and wakame. Kelp is also rich in sodium, potassium, and magnesium. Wakame contains 872 milligrams of sodium, highest among all selected seaweeds.

B-Vitamins in Edible Seaweeds

Edible seaweeds are a good source of B-vitamins. Major B-vitamins present in them are thiamin, niacin, riboflavin, vitamin B-6, and folate. Vitamin B12 is not naturally present in most of the edible seaweeds.

Vitamin B Composition of Seaweeds Per 100g Edible Portion (unit in milligrams)

Seaweed	Thiamin	Riboflavin	Niacin	Vitamin B-6	Folate
Gracilaria	0.005	0.022	0.055	0.032	85
Irish Moss	0.015	0.466	0.593	0.069	182
Kelp	0.05	0.15	0.47	-	180
Laver	0.098	0.446	1.47	0.159	146
Spirulina	0.222	0.342	1.196	0.034	9
Wakame	0.06	0.23	1.6	0.002	196

Source: USDA database

Spirulina contains considerable amounts of thiamine, riboflavin and niacin but negligible amounts of folate as compared to other seaweeds. Irish moss, kelp, laver and wakame are rich sources of folate. Wakame is rich in riboflavin and niacin also. Laver is rich in vitamin B6.

Vitamin A, C, E, and K in Edible Seaweeds

Seaweeds are considered to be rich source of all essential vitamins such as vitamin A, B, C, E, and K. Vitamin D is not naturally found in any of the edible seaweeds.

Vitamin A, C, E, K in Seaweeds Per 100g Edible Portion (unit in micrograms)

Seaweed	Vit C	Vit A	Vit K	Vit E
Gracilaria	-	-	2.3	0.87
Irish	3	-	5	0.87
Kelp	3	6	66	0.87
Laver	39	260	4	1
Spirulina	0.9	3	2.5	0.49
Wakame	3	18	5.3	1

Source: USDA database

Laver is richest source of vitamin C and vitamin A among the listed seaweeds. All listed seaweeds are rich in vitamin E. Kelp is the richest source of vitamin K.

Protein, Lipid, Carbohydrate and Dietary Fibre in Seaweeds

Protein, lipid and carbohydrate profile of six popular seaweeds is given below:

Protein, Fat, Carbohydrate, Fiber and Sugars in Edible Seaweeds Per 100g of Edible Portion (unit in grams)

Seaweed	Protein	Fat	Carbohydrate	Fibre	Sugars
Gracilaria spp.	0.54	0.03	6.75	0.5	0.28
Irish moss	1.51	0.16	12.29	1.3	0.61
Kelp	1.68	0.56	9.57	1.3	0.6
Laver	5.81	0.28	5.11	0.3	0.49
Spirulina	5.92	0.39	2.42	0.4	0.3
Wakame	3.03	0.64	9.14	0.5	0.65

Source: USDA database

In 100 grams of fresh, edible portion of spirulina, 5.92 grams of good quality protein is available. Spirulina is considered as one of the highest source of good quality plant protein. All listed seaweeds are rich in dietary fibre and therefore, they are a good choice for a balanced diet. All these seaweeds are low in fat and sugars.

5 RED SEAWEED GRACILARIA

Gracilaria and *Geladium* species of red algae are used as a food in Japanese, Hawaiian, and Filipino cuisine. In Japanese, *Gracilaria* is called ogonori or ogo. Fresh Gracilaria spp. is considered as a salad vegetable in many parts of the world. *Gracilaria confervoides* and *Gracilaria eucheumcides* are used for preparing salad. Clean and washed agar seaweeds are blanched and mixed with chopped onions, tomatoes and ginger to prepare the salad. *Gracilaria eucheumcides* is chopped and added as a vegetable in fish preparations. *Gracilaria edulis* is used to prepare seaweed porridge in the coastal areas of Southern India. There, these seaweeds are collected from the sea shores and cleaned thoroughly before boiling it in water. They are boiled just like rice boiled with water. Cooked seaweeds are taken out and salt is added to the taste before serving it. Seaweed porridge is considered as a highly nutritious meal in these parts of the world.

Red algae are also used for making jelly, extracting agar, and various other medicinal purposes. They are known to have aphrodisiac properties.

Small scale cultivation of agar seaweeds has been picked up by many countries in recent years. *Gracilaria spp.* can be cultivated for food purposes by using several methods. It can be grown vegetatively in open waters on the bottom of bays, estuaries or reef flats. Rope farming by using ropes or nets may also be practiced. Pond cultivation and tank cultivation are also possible. Under well-managed tank cultivation system, a *Gracilaria* crop may produce up to six tons of fresh matter per week.

Nutritional Information
Fresh agar seaweed is rich in minerals such as calcium, potassium and magnesium and also rich in folate vitamins. Nutritional information of both fresh and dried agar seaweeds is given below:

Nutrition in Agar Seaweeds		Raw	Dried
Nutrient	Unit	Value per100g	
Water	g	91.32	8.68
Energy	kcal	26	306
Protein	g	0.54	6.21
Total lipid (fat)	g	0.03	0.3
Carbohydrate, by difference	g	6.75	80.88
Fibre, total dietary	g	0.5	7.7
Sugars, total	g	0.28	2.97
Calcium, Ca	mg	54	625
Iron, Fe	mg	1.86	21.4
Magnesium, Mg	mg	67	770
Phosphorus, P	mg	5	52
Potassium, K	mg	226	1125
Sodium, Na	mg	9	102
Zinc, Zn	mg	0.58	5.8
Thiamin	mg	0.005	0.01
Riboflavin	mg	0.022	0.222
Niacin	mg	0.055	0.202
Vitamin B-6	mg	0.032	0.303
Folate, DFE	µg	85	580
Vitamin E (alpha-tocopherol)	mg	0.87	5
Vitamin K (phylloquinone)	µg	2.3	24.4

Source: USDA Database

Gracilaria and *Geladium* species of red algae are also known as *agar seaweeds* because agar, a solidifying component of bacteriological culture media, is mainly extracted from these seaweeds.

Preparation of Agar Powder

Fresh agar seaweeds are collected. They are washed and cleaned before drying in the sun for three consecutive days. The dried seaweed is ground with water by using a stone mortar. The pulp is, then, spread in a tray with water and left at a cool, well-aerated and dry place for 24 hours. After that, the pulp is dried in the sun thoroughly. Dried pulp is then pulverized by using a grinder. Thus agar powder is obtained. Agar powder is used in small quantities for adding into ice-cream mix, tomato soup, jams, jelly and marmalade to improve the consistency of the final product.

Preparation of Agar Jelly

For preparing agar jelly, one liter of water is boiled in a steel vessel and then, 100 grams of agar powder is added into it. Stir this mixture time to time until the mixture thickens. Now it's time to take the mixture out from the heat and filter it through a strainer or a cloth filter. Add adequate amount of sugar to the filtered jelly to the taste and add other ingredients such as lime juice, essence of your choice and coloring agent. Mix the contents well before filtering the mixture again into a clean tray. Keep the mixture to cool. Agar jelly is now ready to serve. Cut it into small pieces and serve.

Uses of Agar

Gracilaria spp and Galidium spp. are used to extract a carbohydrate called agar-agar. Agar-agar is extensively used in pharmaceutical, cosmetics, and paint manufacturing industry. Agar-agar is also used a culture medium in microbiology and tissue culture laboratories. In food industry, agar is used in canning meat, fish, and poultry. Agar is an effective clarifying agent in brewing and wine making. It is also used as a thickening and stabilizing agent in ice cream, pastries, desserts, and salad dressings.

6 IRISH MOSS

Scientific name of irish moss is *Chondrus crispus* and it belongs to the family gigartinaceae. Irish moss or Carrageenan Moss is small, reddish brown algae, found growing abundantly along the shores of Ireland and Atlantic coast of Europe and North America. Fresh irish moss is soft and mucilaginous in texture. The mucilaginous body of irish moss is mainly made of a polysaccharide called *carrageenan*. Carrageenan makes up to 55% of its weight. Irish moss is used for various food purposes. It is used for jelly making, and as a clarifying agent for beer brewing. In some countries of Europe, irish moss is used for preparing *blancmange*, a sweet dessert made from milk and sugar with irish moss used as a thickening agent.

Irish Moss for Jelly Preparation

Collect the seaweeds and clean it thoroughly in running water. Dry it in the sun and comminuted into small fragments. These seaweed fragments are then boiled with water until it fully dissolves into water and a syrupy consistency is obtained. This syrup is then spread on a large tray and taken to gentle heat until all water is evaporated and large sheets of jelly is remained. Jelly sheets are shredded and stored in clean jars for later uses. Irish moss jelly has wide applications in culinary and medicinal fields as a thickening and stabilizing agent.

Irish Moss as a Clarifying Agent in Home brewing

A small amount of irish moss is added into the kettle along with the brewing wort. All the solids in the brewing mixture are then collected by the irish moss which is then removed from the mixture after cooling.

Irish Moss is an Industrial Source of Carrageenan.

Irish moss seaweeds are used to extract *carrageenan*, a polysaccharide used as a thickener and stabilizing agent in milk and milk products.

Nutritive Value of Irish Moss

Irish moss is one of the richest sources of iron among the vegetarian diets. It also contains almost all B-vitamins, except vitamin B12. It is rich in dietary fibre and low in fat and sugars thus making it suitable for a healthy diet.

Nutrition in Raw Irishmoss Seaweed

Nutrient	Unit	Value per100g
Water	g	81.34
Energy	kcal	49
Protein	g	1.51
Total lipid (fat)	g	0.16
Carbohydrate	g	12.29
Fibre, total dietary	g	1.3
Sugars, total	g	0.61
Calcium, Ca	mg	72
Iron, Fe	mg	8.9
Magnesium, Mg	mg	144
Phosphorus, P	mg	157
Potassium, K	mg	63
Sodium, Na	mg	67
Zinc, Zn	mg	1.95
Vitamin C	mg	3
Thiamin	mg	0.015
Riboflavin	mg	0.466
Niacin	mg	0.593
Vitamin B-6	mg	0.069
Folate, DFE	µg	182
Vitamin A, RAE	µg	6
Vitamin A, IU	IU	118
Vitamin E	mg	0.87
Vitamin K	µg	5

7 WINGED KELP

Kelps are a large group of seaweeds belonging to the brown algae class Phaeophyceae. There are about 30 different genera of kelps, out of which very few are considered edible. In this chapter, we discuss about *Alaria esculenta*, an edible kelp. It is commonly known as winged kelp. Other names of winged kelp are dabberlocks or badderlocks. Winged kelp belongs to the family Alariaceae.

This is a large brown seaweed having a wide distribution in shallow, cold waters and commonly found in coastal areas of Europe, North America and Japan. *Alaria esculenta* is a well-known seaweed in Ireland and it is a perennial algae.

Winged kelps need a temperature between 6°C and 14°C (43°F and 57°F) for their healthy growth. They do not survive above 16°C. Winged kelps have a high growth rate and sometimes it grows to a maximum length of two meters. Its fronds are brown which consists of a distinct midrib with wavy, membranous lamina on either side.

It is eaten raw as a salad after removing its midrib. It is suitable for cooking as a vegetable. Dried kelps are also consumed. Winged kelp is considered as an excellent source of iodine, iron, and other minerals. It is also considered as one of the best sources of the best quality plant protein. Kelp powder and flakes are available in the market as food supplements.

Alaria esculenta is used for alginate production and also as animal fodder. It has wide applications in cosmetics industry especially in the preparation of body care products.

Alaria esculenta is rich in minerals such as calcium, iron, magnesium, phosphorus, potassium, sodium, and zinc as well as vitamins A, C, E, K and B-vitamins thiamine, niacin, riboflavin and folate. Because of the presence of quality vitamins, these seaweeds are believed to increase immunity of human body. A detailed account of nutrient value of *Alaria esculenta* is given below:

Nutrition in Raw Kelp Seaweed

Nutrient	Unit	Value per 100g
Water	g	81.58
Energy	kcal	43
Protein	g	1.68
Total lipid (fat)	g	0.56
Carbohydrate, by difference	g	9.57
Fiber, total dietary	g	1.3
Sugars, total	g	0.6
Calcium, Ca	mg	168
Iron, Fe	mg	2.85
Magnesium, Mg	mg	121
Phosphorus, P	mg	42
Potassium, K	mg	89
Sodium, Na	mg	233
Zinc, Zn	mg	1.23
Vitamin C, total ascorbic acid	mg	3
Thiamin	mg	0.05
Riboflavin	mg	0.15
Niacin	mg	0.47
Folate, DFE	µg	180
Vitamin A, RAE	µg	6
Vitamin A, IU	IU	116
Vitamin E	mg	0.87
Vitamin K	µg	66

8 GREEN LAVER

Green laver is also known as *sea lettuce*. . Scientific name of Green Laver is *Ulva lactuca* and it is one of the highly nutritious seaweeds. Green laver is sometimes referred as 'aonori' or green seaweed. This green algae is now commercially cultivated in some Asian countries including Japan and Taiwan. Dried and powdered laver seaweeds are used in soups and as a flavouring agent in some Japanese food preparations such as fried noodles, Japanese pancake, and Japanese potato chips. For maximum flavouring and seasoning effect, the aromatic powder is sprinkled on the hot food after serving. Laver is a favourite among the Welsh people also and there it is commonly used for making laver bread, a traditional Welsh delicacy.

Preparation of Laver Bread

Fresh laver is collected, cleaned well under running water. Clean laver is placed in a vessel and add adequate amount of water and boil it for several hours until it is cooked well. Cooked laver is then made into a puree. The gelatinous paste of laver is then rolled in oatmeal and fried to make laver bread.

Laver as a Side Dish

Cooked laver is used as a side dish along with meat preparations such as mutton and bacon. For cooking laver, fresh laver seaweeds are cleaned thoroughly before placing it in a vessel and heat it. While heating, add adequate amounts of butter and lime juice or orange juice. Mix it well and serve hot.

Nutrition in Green Laver

Green Laver seaweeds are a group of green algae that is rich in iodine and iron. It contains easily digestible protein and considerable amounts of dietary fibre. It is low in sugars, rich in minerals particularly, calcium, phosphorus, potassium and sodium. It is an excellent source of vitamin A, vitamin C and folate. Nutritional information of raw laver is given below in detail:

Nutrition in Green Laver

Nutrient	Unit	Value per100g
Water	g	85.03
Energy	kcal	35
Protein	g	5.81
Total lipid (fat)	g	0.28
Carbohydrate, by difference	g	5.11
Fibre, total dietary	g	0.3
Sugars, total	g	0.49
Calcium, Ca	mg	70
Iron, Fe	mg	1.8
Magnesium, Mg	mg	2
Phosphorus, P	mg	58
Potassium, K	mg	356
Sodium, Na	mg	48
Zinc, Zn	mg	1.05
Vitamin C, total ascorbic acid	mg	39
Thiamin	mg	0.098
Riboflavin	mg	0.446
Niacin	mg	1.47
Vitamin B-6	mg	0.159
Folate, DFE	µg	146
Vitamin A, RAE	µg	260
Vitamin A, IU	IU	5202
Vitamin E (alpha-tocopherol)	mg	1
Vitamin K (phylloquinone)	µg	4
Fatty acids, total saturated	g	0.061
Fatty acids, total monounsaturated	g	0.025
Fatty acids, total polyunsaturated	g	0.11

9 PURPLE LAVER

Nori (*Porphyra spp.*) or Purple Laver includes a large number of species such as Porphyra yezoensis, P. tenera, P. umbilicalis and Porphyra spp. However, the original Nori is *Porphyra yezoensis*. This belongs to the family Bangiophyceae. Nori is deep purplish-coloured edible seaweed of temperate regions. It is also known as purple laver because of its purple colour. This seaweed is normally seen on growing as a thin layer on rock surfaces or other hard substrate surfaces along the sea shores.

Nutritional Value of Nori

Nutritional value of nori is very high. It is rich in three essential amino acids such as alanine, glutamic acid and glycine. Due to the presence of glutamic acid, nori can be used as a food additive to enhance the taste of the food preparations. Nori is rich in provitamin A. Dried nori contains large amounts of protein, ash, vitamins and carbohydrate. The fat present in nori is of great nutritional value as more than 60% of them are omega 3 and omega 6 polyunsaturated fatty acids.

Food Value of Nori

Nori is considered as a luxury food in Japan. Nori is familiar in many countries as a major ingredient in sushi preparations. Nori is used as a flavouring agent in noodle and soup preparations.

Preparation of Nori for Food Purposes

Freshly harvested fronds of nori are chopped, and then pressed between two bamboo mats for drying. Nori can be dried either in the sun or in the drying rooms. Properly dried nori will have a purplish black color. Dried flakes of nori are stored in air tight containers as it is hygroscopic (i.e. absorbs moisture). High quality Nori is mild-tasting. These purplish-black nori sheets are used to wrap around rice in sushi cuisines of Japan.

10 SPIRULINA

Spirulina is one of the most talked about super foods today. Spirulina is actually a blue-green aquatic cyanobacterium or a common form of blue-green algae with photosynthetic capabilities. There are about 60 different species of spirulina. However two of them are commercially cultivated. Scientific name of cultivated species of spirulina are *Arthrospira (Spirulina) platensis* and *Arthrospira (Spirulina) maxima*. *Arthrospira maxima* is found growing in Central American countries while *Arthrospira platensis* is common in Asian and African countries. Spirulina is commercially produced in many Asian countries including Thailand, India, Taiwan, China, Bangladesh, Pakistan, and Myanmar. It is produced in the United States of America also in considerable quantities. Other major producers are Greece, and Chile. Spirulina thrives well in warm, alkaline fresh waters. It requires a pH of 8.5 or above for its healthy growth. Optimum temperature for its growth is 30 °C (86 °F). Spirulina is suitable for tank cultivation and pond culture.

Spirulina is rich in pigments and it contains three pigments namely, chlorophyll, phycocyanin, and beta-carotene that give it a green, blue and orange tints respectively. Chlorophyll helps this blue-green algae for photosynthesis by which it synthesizes its own food. Pycocyanin is a valuable antioxidant. Beta-carotene is precursor to vitamin A. Production of spirulina as a protein-rich food has many advantages. It has very high protein content, up to 70% of dry weight. Therefore spirulina seaweeds are used as protein supplements. Top quality spirulina supplement has a solid, dark green colour, and it quickly dissolves into water. Spirulina protein is easily digestible and its protein quality is one of the best in the plant world. Any amino acid imbalance in a diet can be easily corrected by consuming spirulina-rich protein food. In other words, spirulina microorganisms represent one of the richest protein sources of plant origin. Spirulina food supplements are available in the market as tablet, flake and powder forms. Spirulina biomass is also used in aquaculture and poultry industries as feed supplements. Spirulina is known for its anticancer and antitumor properties. It is also considered as an antiviral, antibacterial and antifungal agent.

Nutrition in Spirulina Seaweed

Nutrient	Unit	Raw	Dried
		Value per 100g	
Water	g	90.67	4.68
Energy	kcal	26	290
Protein	g	5.92	57.47
Total lipid (fat)	g	0.39	7.72
Carbohydrate, by difference	g	2.42	23.9
Fibre, total dietary	g	0.4	3.6
Sugars, total	g	0.3	3.1
Calcium, Ca	mg	12	120
Iron, Fe	mg	2.79	28.5
Magnesium, Mg	mg	19	195
Phosphorus, P	mg	11	118
Potassium, K	mg	127	1363
Sodium, Na	mg	98	1048
Zinc, Zn	mg	0.2	2
Vitamin C, total ascorbic acid	mg	0.9	10.1
Thiamin	mg	0.222	2.38
Riboflavin	mg	0.342	3.67
Niacin	mg	1.196	12.82
Vitamin B-6	mg	0.034	0.364
Folate, DFE	µg	9	94
Vitamin A, RAE	µg	3	29
Vitamin A, IU	IU	56	570
Vitamin E (alpha-tocopherol)	mg	0.49	5
Vitamin K (phylloquinone)	µg	2.5	25.5
Fatty acids, total saturated	g	0.135	2.65
Fatty acids, total monounsaturated	g	0.034	0.675
Fatty acids, total polyunsaturated	g	0.106	2.08

Dried spirulina contains 57.47 % protein and it contains all essential amino acids thus making it a complete protein food. It is a rich source of dietary fibre, minerals and vitamins, thus making it a most-sought after super food. Spirulina does not contain vitamin B-12 naturally and hence not recommended as a Vitamin B-12 source.

11 WAKAME

Scientific name of Wakame or Quandai-cai is *Undaria pinnatifida* and it belongs to the family Alariaceae. Sometimes it is referred as '*sea mustard*'. Wakame is believed to be a native to the Japan Sea. It is brown-coloured edible algae grown as an annual seaweed of temperate climate. It is naturally found growing on the surfaces of rocks or other hard substrates along the sea shores. It is invasive in nature. These seaweeds are found growing in deep waters. Wakame is one of the most important species of commercial, edible seaweed, next to nori, in Japan. Wakame is commercially cultivated in Japan, China and Korea in large scales. It is also cultivated in France, Tasmania and New Zealand.

Nutritional Significance of Wakame

Wakame is rich in vitamin B group; it has also considerable amounts of manganese, copper, cobalt, iron, nickel and zinc. Wakame is a rich source of good quality protein and calcium. It is high in dietary fiber and therefore has many health benefits as it reduces the risk of colon cancer, removes constipation, and good for treating hypercholesterolemia, obesity and diabetes. The fat content is quite low in wakame. Vitamin content in both fresh and air-dried wakame is almost same. However, when wakame is processed, all its vitamins are lost. Wakame contains 5–10% fucoxanthin, an important antioxidant. *Fucoxanthin* is is a characteristic carotenoid of brown algae and is believed to have anti-obesity effects in diet-induced obesity. Researches on various pharmacological properties of fucoxanthin showed that fucoxanthin has fat-burning properties. Fucoxanthin has also shown a great antioxidant activity, and anti-cancer properties. It is believed to have anti-diabetic and anti-aging properties as well.

Wakame is also used for its anti-inflammatory activity. Wakame is a rich source of eicosapentaenoic acid, an omega-3 fatty acid. It has high levels of sodium, calcium, iodine and magnesium. Vitamin B12 is not naturally present in wakame and presence of Vitamin B6 is also negligible in wakame. Wakame is good for blood purification, and intestinal strength. It is used in skin and hair beauty treatments. Nutritional value of 100 grams of edible portion of fresh wakame is given below:

Nutrition in Raw Wakame Seaweed

Nutrient	Unit	Value per100g
Water	g	79.99
Energy	kcal	45
Protein	g	3.03
Total lipid (fat)	g	0.64
Carbohydrate, by difference	g	9.14
Fibre, total dietary	g	0.5
Sugars, total	g	0.65
Calcium, Ca	mg	150
Iron, Fe	mg	2.18
Magnesium, Mg	mg	107
Phosphorus, P	mg	80
Potassium, K	mg	50
Sodium, Na	mg	872
Zinc, Zn	mg	0.38
Vitamin C, total ascorbic acid	mg	3
Thiamin	mg	0.06
Riboflavin	mg	0.23
Niacin	mg	1.6
Folate, DFE	µg	196
Vitamin A, RAE	µg	18
Vitamin A, IU	IU	360
Vitamin E (alpha-tocopherol)	mg	1
Vitamin K (phylloquinone)	µg	5.3
Fatty acids, total saturated	g	0.13
Fatty acids, total monounsaturated	g	0.058
Fatty acids, total polyunsaturated	g	0.218

Wakame Products

Wakame is consumed in dried, frozen and fresh forms. It has a subtle sweet flavour and satiny texture. Wakame is marketed as the frozen sporophylls in Japan and other countries. Sporophylls are the flowering sprouts located at the base of *Undaria pinnatifida* just above the root. Sporophylls of brown algae such as *Undaria pinnatifida* is known to be a source of Fucoidan. Fucoidan (fucan sulphate) is a fucose-containing sulphated polysaccharide. Fucoidan is believed to be having anticoagulant and anti viral properties. In Japan, the frozen sporophylls of *Undaria pinnatifida* is known as *mekabu*, a savoury food. Wakame is also marketed as cooked and salted leaf blades and midribs. In Japan markets, dehydrated or dried, seasoned and instant wakame products are also available.

Food Uses of Wakame

Wakame is commonly used in broths, soups, pickles and salads. Dried powder of wakame is used as a garnishing agent in fried noodles and rice preparations. Raw wakame leaves (fronds) are used as vegetables. Since wakame fronds enlarge while cooking, it is advised that chopped fragments are used for cooking it.

For using as vegetables, collect wakame and remove the roots and upper leaves. Clean it thoroughly under running water. Immerse clean and fresh wakame in boiling water for 30 seconds. Take it out from boiling water and rinse with ice-cold water. Spread out cooked wakame in a tray and remove hard midribs one by one. Then chop the leaves into small fragments and mix it well with other salad ingredients. Wakame salad is ready.

12 DULSE

Scientific name of Dulse or Dilisk is *Palmaria palmate* and it belongs to the family Florideophyceae. Dulse is a small, reddish Atlantic seaweed found growing naturally along the coastal areas of Europe (Scotland, Ireland, and Iceland) and in North America (Canada and the United States of America).Today, dulse is successfully cultivated along the coastal areas of France, Ireland and Spain. Dulse is the seaweed marketed under the trade name '*Sea Parsley*' in Canada. Dulse is actually a red-coloured algae having perennial growth habit. Leaves (fronds) are leathery in appearance. Harvesting is done during May to October. Normally manual harvesting is done by hand picking. Shelf life of fresh dulse is very less and therefore processing by sun drying is recommended in order to extend its shelf life. As a standard, fresh dulse is sun-dried for six to eight hours to dehydrate it. Clean and dried dulse is then packed in air-tight plastic bags.

Fresh dulse is eaten raw as a salad component in Ireland and surrounding areas. In Iceland, dulse is eaten with butter. Fresh dulse is used in soups also. Fresh dulse can be pan-fried into chips. It can also be baked in the oven covered with cheese. Fresh dulse is chopped into fine fragments and used in meat preparations to enhance taste. The presence of glutamic acid in dulse increases the taste of food preparations and hence it can be used as a food additive. Dehydrated (sun-dried) dulse is eaten uncooked as a snack. Dried dulse is ground to flakes or a powder and is used as a food seasoning agent. Dried dulse is a popular food in Canada which is served as a side dish, in soups and salads, as a sandwich component or as a seasoning powder. Nutritive value of dulse is very high. Dulse is a good source of dietary fibre. Dulse contains all trace elements needed by humans. It is rich in minerals and vitamins. Dulse is high in minerals such as iodine, phosphorus, calcium, and potassium and it is estimated that about 30% of the dry weight of dulse is made up of minerals. Dulse consists of proteins of high nutritive value (about 18%) and high amounts of vitamin C. Vitamin C plays a crucial role in iron absorption. Hence dulse diet is recommended for anemic patients. Dulse also has anthelmintic effect and acts as an antiseptic. Since dulse is rich in iodine, it can be used to prevent goiter.

13 CAULERPA

Many species of caulerpa seaweeds can be found in the sea. However, *Caulerpa lentillifera* and *C. racemosa* are the two most popular edible seaweeds. They belong to the family Caulerpaceae. Caulerpa seaweeds are grape-like in appearance due to their grass-green colour. They are soft and succulent. These grape-like edible algae have a peppery taste and are used as salad vegetable. It can be consumed both as raw or cooked form. They are also known as *'Sea Grapes'* or *Green Caviar*. Caulerpa is suitable for pond culture. They are naturally found on sea bottoms in sub-tropical areas.

Nutritional Information

Protein content of *C. lentillifera* is 12.49%. Caulerpa is low in sugars and fat, making it a suitable food for weight loss. It is rich in minerals such as iodine, phosphorus, calcium, copper and magnesium. It is also rich in vitamin E with moderate amount of vitamin B1, vitamin B2 and niacin.

Food Uses

Caulerpa racemosa as a Salad: These seaweeds are collected and washed. After thorough cleaning, they are mixed with chopped onions and tomatoes. Salt is added to taste. Seaweed salad is ready.

Caulerpa serrulata as a Vegetable: Seaweeds are washed and cleaned thoroughly in running water. Clean seaweeds are then taken in a vessel and boiled water is added into it until the seaweeds are immersed in it. Keep it aside for one minute before draining the water. Dry seaweeds are then mixed with chopped onions, ginger and tomatoes. Finally fermented fish sauce (patis) is poured into it and the mixture is mixed well before serving. Salt is added to the taste.

14 KOMBU

Scientific name of Kombu or Haidai is *Saccharina japonica* (formerly known as Laminaria japonica) and it belongs to the family Laminariaceae. Kombu is an edible kelp and is widely eaten in East Asian countries. Kombu has broad, shiny leaves and grows naturally in cool waters off the coasts of Japan and Korea. It is cultivated on a large scale in China. Kombu is a large, brown seaweed of tropical climate and is one of the most consumed algae around the world. In pond culture, kombu is grown as an annual seaweed while naturally growing kombu is a biennial seaweed.

Health Benefits of Kombu

Kombu is rich in protein (10 %) and low in fat (2%) and sugars. Hence cconsumption of kombu is believed to regulate body weight. It contains considerable amounts of minerals and vitamins. Kombu is high in magnesium, calcium and iodine. Kombu seaweeds are anti-rheumatic, and anti-inflammatory. It is good for controlling blood pressure due to the presence of bioactive compounds, laminarin and laminin.

Food Uses of Kombu

Kombu is used for a variety of food purposes. Kombu is eaten raw as salads, and can be pickled. Dried and powdered kombu is used as seasoning and food flavoring agents in broths, stews and soups. It is used as a garnishing agent for rice preparations such as sushi. Dried shreds of kombu are used for food decorative purposes. Kombu is also consumed as a vegetable and as snacks. Kombu when adding to cooking beans improves its digestibility by converting the indigestible sugars to digestible form. Kombu is also rich in glutamic acid, an amino acid that gives taste to food.

15 GROWING SEAWEEDS

Seaweeds are a group of marine algae that are grown in water. So the most essential requirement for seaweed cultivation is good quality water. Depending on the type of seaweeds, their cultivation requirements may differ up to some extent. There are three types of seaweeds – red algae (Rhodophyta), brown algae (Phaeophyta), and green algae (Chlorophyta). Green algae contain chlorophylls and require sun light for food synthesis. So for growing green seaweeds, sunny location may be preferred. Open cultivation is practiced for growing seaweeds on a large scale. Seaweed farming is common in many Southeast Asian countries, Canada, Spain, some European countries and the United States of America. Different methods of cultivation are practiced in different countries. Hang method of cultivation where seaweed cuttings are hanged into the water from a bamboo plot fixed above the seaweed farm is common in some South East Asian countries. Generally, net farming and rope farming methods of seaweed cultivation is practiced commercially.

Seaweeds may be grown naturally in shallow or deep water fields situated near the sea shores. Major advantage of this cultivation method is that naturally-existing seawater nutrients may be made available to the growing seaweeds by pumping the seawater into the farm. Seaweeds may also be grown in artificially created ponds or tanks. In fact, pond culture and tank cultivation are practiced for small scale or home scale growing of seaweeds. A detailed account of seaweed farming practices is given below:

Selection of Farm Site

Seaweeds are delicate plant-like organisms and therefore, site for seaweed farming should be selected according to the nature of the seaweeds. Seaweed farm should have good quality water (with required water salinity), good water movement, and required depth of water. Saline sea water is the best suited for seaweed growing. Some seaweeds may be grown by using fresh water also. Optimum water temperature is between 25°C and 30°C. The site should be protected from strong winds and water currents. Areas with heavy water currents are not suitable for seaweed farming. Farm bottom should be clean, and free of other sea vegetation. Strong, stable, sandy farm bottom is preferred for seaweed cultivation.

Construction of Seaweed Farm by Using Nets

For constructing a small unit, first of all, four wooden stakes are installed into the farm bottom at the four corners of the farm. A polyethylene net of appropriate size is stretched tightly and each corner of which is tied to each of the stakes by using plastic ties. Net is fixed at least two feet above from the farm bottom horizontally. For constructing a large farm, such small units are replicated all over the site.

Construction of Seaweed Farm by Using Durable Ropes

Rows of wooden stakes are installed into the farm bottom at convenient length. Recommended spacing is one meter within the rows and 10 meter between the rows. Durable polyethylene ropes are tied to both ends of the wooden stakes. The rope should be fixed at least one meter above the farm bottom.

Planting Seaweed Cuttings

Healthy cuttings are taken from the central part or the base of the mother seaweed. After cleaning the cuttings, they are introduced into farm by tying them on the nets/ropes with soft plastic ties. One cutting is planted at each planting site of the ropes/nets. In net method, each intersection of the net may be used as a planting site. In rope method, 20–25 cm space is left between two cuttings on the rope. After introducing the cuttings into the seaweed farm, they are left undisturbed to grow until ready to harvest. Seaweeds grow at a rapid rate and first crop can be harvested within two months of introducing the cuttings into the farm.

Harvesting

Multiple harvests are possible. Surface canopy of a full grown seaweed farm is harvested by using boats and seaweed-pulling equipments. Basal portion of the seaweeds is left for regrowth. Harvested seaweeds are cleaned thoroughly and dried in the sun for 2 to 3 days (depending on the climate). Spread the wet seaweeds in a clean, hygienic drying site to avoid contamination.

16 HEALTH BENEFITS OF SEAWEEDS

Seaweeds are now considered as one of the healthiest foods that contribute towards health and wellness of consumers. Some of the most important health benefits of seaweeds are briefed below:

Seaweeds as Protein Supplements: Seaweeds are excellent sources of good quality protein which is easily digestible and available to the human body.

Seaweeds for Blood Clotting, Blood Purification and Immunity: Seaweeds are high in iron, which increases body immunity. Certain bioactive compounds present in seaweeds believed to have blood purifying properties. Seaweeds are high in Vitamin K, which is essential for blood clotting.

Seaweeds Promote Bone Health: Seaweeds are high in calcium, almost ten times more calcium than that is present in milk.

Seaweeds Regulate Blood Pressure: Seaweeds are high in magnesium which is essential for regulating blood pressure. Magnesium also promotes kidney health.

Seaweeds Have Alkalinizing Effect on Human Body: Seaweeds are high in sodium; so it alkalinizes human bodies and thus reduces the adverse effects of acidity.

Seaweeds Have Anti-Cancer Effect: Seaweeds are rich in antioxidants that oxidize cancer-causing free radicals into harmless molecules.

Seaweeds Have Detoxifying Properties: Seaweeds contain moderate amounts of dietary fibers which play a crucial role in eliminating toxins from the human body.

Seaweed Consumption Prevent Goiter: Goiter is a disorder caused by iodine deficiency. Seaweeds are rich in iodine and hence its regular consumption reduces the incidence of goiter. Iodine is also an antioxidant mineral. Iodine plays a crucial role in boosting natural body metabolism.

Seaweeds Have Anti-Obesity and Weight Loss Properties: Seaweeds are rich in protein but low in fat and sugars. Seaweeds are also low in carbohydrates thus making it ideal for a low-carb diet. Regular seaweed consumption is believed to regulate body weight and cut down obesity.

Seaweeds Promote Health: Seaweeds are high in Vitamin A which is essential for promoting skin and hair health. Vitamin E is also present in seaweeds, which is a powerful antioxidant vitamin. Seaweeds are excellent source of folate, riboflavin and other B-vitamins which are necessary for promoting reproductive health.

Seaweeds for Preventing Heart Diseases: Seaweeds are rich in Vitamin K, which is essential for preventing heart diseases, cancer, and osteoporosis.

Seaweeds for Treating Arthritis: Seaweed baths are used in some Asian countries for treating arthritis and rheumatism.

BIBLIOGRAPHY

Food and Agriculture Organization. (2014, July Tuesday). Retrieved July Tuesday, 2014, from http://www.fao.org/docrep/006/y4765e/y4765e0b.htm

Germplasm Resources Information Network . (2014, October Monday). Retrieved October Monday, 2014, from GRIN NPGS: http://www.ars-grin.gov/npgs/

Handbook of Agriculture. (2005). New Delhi: ICAR.

UC Davis Post Harvest. (2014, October Monday). Retrieved October Monday, 2014, from UC Davis Post Harvest Technology Center: http://postharvest.ucdavis.edu/producefacts/

USDA ARS . (2014, October Monday). Retrieved October Monday, 2014, from USDA Agricultural Research Service: http://www.ars.usda.gov/main/main.htm

USDA Nutrient Database . (2014, October Monday). Retrieved October Monday, 2014, from USDA Nutrient Database: http://ndb.nal.usda.gov/ndb/search/list

USDA Plant Database . (2014, October Monday). Retrieved October Monday, 2014, from USDA Plant Database: http://plants.usda.gov/java/

ABOUT THE AUTHOR

Roby Jose Ciju is the author of *'The Art of Perfect Living'*, an inspirational book based on scriptural wisdom. She is a professional horticulturist and an agribusiness consultant with a Masters Degree in Horticulture and a Post Graduate Diploma in Agri-Supply Chain Management. She has founded www.agrihortico.com, a website dedicated for publishing information on Food & Agriculture Topics. She has written more than 40 books on various horticultural topics till date and her best-selling books are, Mushroom Farming, Moringa, Curryleaf, and Growing Ginger, Turmeric and Arrowroot. She may be contacted at roby@agrihortico.com.